The Zata Stone

By

Saaim Khan

Bob Davis had one desire. And it was to become insanely popular. In fact, at the moment he was on his SuperSonic PC in his room. He was searching on how to become (as we know) insanely popular.

"Boring Boring Boring," Bob said as he flipped casually through the suggestions. Suddenly, just as he was about to exit out he found a suggestion that caught his eye.
"The Zata Stone" he read slowly but his voice was filled with anticipation. He had heard many tales of how the

combination of mustard seed and the Elixir of Popularity could make someone extremely popular no matter how they were before. "Is located in Celebrium, a black hole, in the Yoti Gulik, a solar system, in the Tredy Galaxy and there are only two portals, one for the Tredy Galaxy and one for its twin, The Trudy Galaxy." Bob wanted to read more but he already had enough info. He needed to get a CoronaGGTR, the newest at the Chitty Chitty Bang Bang Spaceship Factory. It was capable of traveling 3 trillion light years a minute.

Also, its seats were made of extra soft zunko fabric, the softest fabric in the world. It even had a gravity machine. And it only cost 5 billion dollars! (Considering the fact that his grandma was a double-triple-quadruple-quintuple-multi-centi billionaire, in other words really rich!). Although it was a magnificent machine it needed 1.21 gigawatts of electricity. But plutonium could take care of that.

"Bob," his grandma called. "Time for dinner."

Bob groaned and slammed his head over the keyboard. He had forgotten that today was Friday. The day when Stumpy joined them for dinner. Stumpy was a wad of money from every country in the world--- literally. He was worth so much that his grandma brought him to dinner on Fridays. His grandma would act so sweet and kind to Stumpy (even though he couldn't talk) while Bob was treated like a bad boy. He always wanted to rip Stumpy bill-to-bill, dollar-to-dollar. But again Stumpy was worth so much that Bob kept his hands to himself. And anyway, since Stumpy lived in his grandmother's gigantic-as-big-as-a-pantry lock box he couldn't access the wad of money. Today Bob needed his grandma's permission. They had a conversation and this is how it went:

"Grandma"

"Yes?"

"Can I get the CoronaGGTR from Chitty Chitty Bang Bang's?"

"Why?"

"Duh! We'll get the Zata Stone and become popular then rich!"

Immediately his grandma looked curious. Bob knew that catch her attention. She was as greedy for money as a bat is for bugs.

"How much does the newest cost?" she inquired with interest.

"A mere 5 billion" he answered coolly.

"We're going tomorrow." She announced excitedly.

In his room, Bob was buzzing with excitement. Finally he could be more popular than Ching-Chong. Ching-Chong was the most popular kid in Bob's school. He was smart, athletic, strong, fast, and he practically played every sport known to mankind. His schedule was so busy that his mom had to come and tell the teachers that her son couldn't do homework. Bob on the other hand, was fairly smart but *not* athletic, strong, or fast. And he wasn't even close. Hey, he couldn't even do a push-up! But in a few days everything would change. However, something *always* has to go wrong.

-II-

"Bob! *Wake Up*!" his grandma screamed from downstairs. He awoke with a start and sweat was dripping down his forehead. He had forgotten to set his alarm clock. Impetuously, he threw on a pair of jeans and a t-shirt. He rushed down the stairs like a cheetah. His grandma was waiting downstairs nibbling on the tray of her breakfast.

"Just *what* are you doing?" Bob asked. His grandma ignored him. She was wearing her typical outfit. A fancy red velvet shirt decorated with diamond studs and matching pants.

"C'mon, I'm getting hot in this ridiculous outfit" she whined impatiently.

"Then why are you wearing it" Bob asked as he was putting his shoes on. He replaced his laugh with a cough.

"It was the best I had," she protested.

"Let's just *go*" Bob said. They got in their Zunchesko Minko 500 and drove off without a word. They reached in about 5 minutes. When they got there Bob couldn't help staring in awe. Inside, Bob couldn't help but to just say "WHOA!!"

But most of all, he was surprised to see Ching-Chong grinning behind the Customer Service. The idea of a tall, skinny, muscular, pale boy sitting behind a customer service desk when he could be playing bocce ball or lacrosse or even curling amused Bob. Just out of curiosity, Bob walked up and said, "What are you doing here?"

"Dude, my dad owns this sweet shop" Ching-Chong replied.

"Oh" Bob said as he walked away.

When Bob found an employee he asked if he could get the CoronaGGTR. The guy said to pay at the front desk and they would have it out in half an hour. At last they got it in their gigantic garage. Bob was so excited but he knew that he would have competition. Although popular, Ching-Chong was disliked by some people like Bob. And now he wanted those people to like him too. Nevertheless he called his friend Joe Lautrec over. Joe was a *very* smart person. Although he was only ten he had the skills of Albert Einstein. Joe couldn't wait to see Bobo's new spaceship. But as soon as he entered he smelled trouble.

-III-

The problem was that there was an overload in the plutonium tank. That was going to be epic. But Joe sniffed out the overload and they very carefully removed the extra container and put it in their giant box of plutonium using suits that were nuclear-proof and radioactive-proof that were included so that they could change the chamber without dying a painful death. In a few days they were ready. They had 6 million but they found a credit card lying on the ground right before takeoff. When

Bob scanned it in the ship's foreign money scanner they found a happy surprise. The number was astonishing: £∞. Yep, that's right, infinity pounds. They said hasty farewells then boarded the spaceship cleaving to the credit card so tightly that it almost broke. After a few minutes on board came Joe's countdown: 10 9 8 7 6 5 4 3 2 1 BLASTOFF! They rocketed to the sky. All around them it was orange. But Bob was too sick to admire the view. Joe however was perfectly fine. He enjoyed naming particular comets that passed them. Soon they entered the solar system. The pictures that Bob had seen online were nothing compared to the real thing. He saw different colors of nebulae and some stars. They were so wrapped up in the solar system's beauty that they didn't realize that they had left the plutonium.

-IV-

It wasn't until they started to slow down did they realize that they had forgotten the plutonium. Here's how it all started:

"Hey Bob, go get the plutonium. We're running low."

"Where is it?" asked Bob.

"In the back." answered Joe.

"I don't see it," remarked Bob.

"Stop messing with me," laughed Joe.

"I'm not. Come and look for yourself." Bob said.

"Okay" said Joe.

"Hmmm…. I know we put it in here" Joe said. Suddenly his jaw went slack.

"We for…*forgot* it" Joe stammered.

"But on the bright side the next stop is only 5 minutes away" Bob said trying to cheer up Joe.

"According to our navigator it is called the Jumbo Space Place." Joe reported. So as, Bob promised they reached in 5 minutes. When they got there they were astounded and amazed. It made Chitty Chitty Bang Bang's look like an ant compared to this place. They entered and they couldn't resist but to explore the whole entire shop. They were so accommodated in their world of bliss that they forgot about their destiny of becoming popular.

"Finally I can have a head start over that weakling, Bob," Ching-Chong thought grinning. He saw that they had parked over by the Space Place and it didn't look like they were coming out. He was feeling tired so he decided to crash on his miniature bed. But he realized that he had forgotten his *book*. The reason why his book was so important is because it was the only thing that could put him to sleep. And if he didn't sleep then he would just sit there and refuse to do anything. And if he didn't do anything he would just be floating there. So he had to go in the Space Place too. Just like Bob, he couldn't resist the temptation of exploring the whole shop. But as he entered Bobo and Joe saw Ching-Chong and remembered their journey and dream. They checked out 15 giant boxes of plutonium. They immediately rushed onwards and flew to the Tredy Galaxy. On the way they passed Andromeda. They decided to pitch in. It was *very* different from the Milky Way. There were stars a thousand times bigger than the sun. There were even extraterrestrials. They looked like giant ugly ogres. They were wearing huge ugly suits made of something that looked bearskin. Bob and Joe decide to dock at the king's palace and ask for hospitality. Although everything seemed perfect there has to be a flaw.

-VI-

They sent a messenger to the palace saying that they had come in peace. They left the space ship heavily guarded by the ogres. They had something that looked like a rocket launcher and a crossbow. If the ogres were ugly the king was ten times worse. His face was covered in mud and dirt and he had something like a hyena coat on.

"WELCOME, MY FRIENDS FROM OUTER SPACE1" the king's voice boomed. Bob and Joe flinched.

"Hey man, could you take it easy. We're right here," Bob said.

"I AM NOT A MAN!" the king yelled even more loudly than before, "I AM THE KING OF ZUNCHES!"

"Oh" Bob replied, "By any chance, do you have food?"

"YOU ARE VERY LUCKY BECAUSE WE ARE GOING TO HAVE THE FEAST OF YUBIT IN PRECISELY, NOW!" He said that last word so loud that Bobo began wondering why he wasn't deaf. The king's guards nobly escorted them to the mess hall. It was pandemonium. There were mothers, fathers, babies, children, and even some grammas, and grandpas. And it looked like they were getting ready for the feast of Yubit. There were sandwiches being put out and table cloths being laid out. You probably thought of a nice neat room. Wrong! The babies were crying everybody was wearing filthy and dirty clothes and there were kids breaking the dishware. But strangely, no one seemed to care. They all went about their business. Bob and Joe got seats right beside the king, which was supposedly an honor. It wasn't until the king made the announcement, did the Zunches (or ogres---whatever you want to call them) notice them. Some started gawking right away but mostly they were confused and befuddled. But outside the Zunches were so busy playing with each other that they didn't notice a particular person board the spaceship.

That person was Ching-Chong. He had followed Bob to the Andromeda. He also saw them enter the king's palace and he seized his chance to get on the spaceship. He knew that there was a ballista on the spaceship. It was his idea. Here was his plan:

I.   Get on the spaceship

II. Fire the ballista

III. Go back on ship

IV. Go get the Zata Stone

V.   Go home

VI. Eat a chocolate cake

The whole purpose of II and I was so that Bobo and Joe would be mistrusted by the king and eventually held off. So Ching-Chong fired the ballista and eventually the castle crumbled. He ran to his own spaceship and flew off. Out of the rearview mirror he saw Bob and Joe running for their lives while the Zunches were trapped inside the crumbling building. But he didn't look where he was going and crashed.

He crashed into a giant SuperGubbley 500. It was a hundred times bigger than the biggest blue whale on the Earth. It was equipped with a ballista, a laser gun, a bazooka, and a rotating machine gun. It even had a hot tub. It loomed over Ching-Chong's ship like an elephant over an ant. Ching-Chong tried to swerve out of the way but he crashed over the side of the supersized spaceship. Although his ship was like an ant compared to the monstrosity Ching-Chong' s space ship just bounced off harmlessly. And he flew on. In the meantime, Joe and Bobo were trying to find a way to get on their spaceship. Before, Zunches guarded it. Now it was being guarded by angry Zunches. They were pounding their giant weapons in their hands. Their faces read only one clear expression: *Hate*.

"I'll go and make a diversion," Joe yelled. "When I say GO! Run to the ship."

He ran off and did some pretty amazing moves. He did a spinning kick in midair and knocked all the Zunches senseless except for one.

"GO!!!" Joe yelled to Bob. Bob scrambled up to the entrance and got inside but Joe couldn't keep staring. He had an exquisite Zunch to defeat. He was the biggest and the meanest, obviously the leader of the pack. He was wearing a goatskin cloak and his face was covered in soot and mud. He bared his teeth and that was probably the worst sight that Joe saw for a *very long* time. There were flies buzzing around the otherwise black teeth. His gums were a poisonous green. And the guy's weapon---whoa it was a rotating bazooka\machine gun. It was totally beast. Anyway, the guy fired and Joe did a double back somersault\twist kicked him squarely in the chest and stole the guy's weapon. But that weapon- it weighed like a 100 tons! Joe had to operate it on the ground. Luckily, it had wheels! Anyway, Joe blasted him smack dab in the middle of his big gruesome face.

"Beat that you…" Joe started but he didn't finish. The giant boy was rising, his face a hundred times worse than before. There were cracks and even blobs of purple goo where he had been hit. *Blood* thought Joe.

The giant took his own weapon and smashed it to bits in fury and rage.

"Well that was dumb" Joe muttered. But now he had no weapon to face the Zunch.

Suddenly out of nowhere their spaceship came flying in. It repeatedly shot the giant until Joe was sure that he was dead. But he was stunned. Then a huge smile crept up his face. "Bob" he thought. He ran to the spaceship and hugged Bob.

"We ha...have to g...go" Bob managed to choke out. Unfortunately, he was right. According to their navigator Ching-Chong was exactly 60 trillion light years ahead of them. They put the thrust to full and they were on their way. On the way, they saw Ching-Chong's space ship. He saw them, too. He grinned as they passed by and he smashed into them. At first, the collision was weak and Bob didn't notice. Then Ching-Chong kept ramming into them. He didn't know what was making him do it. He felt angry mad and furious at Bob for competing against him. Did that weakling really think that he could compete against one of the fittest kids in school? Although those were his real feelings he felt somehow possessed.

"Ghoujutry" he recalled, "Possessive Spirits." They could be good or bad. In this case they were obviously bad. Anyway, Ching-Chong kept ramming into Bob's spaceship. Finally, Bob started noticing. He and Joe felt a tremor every five seconds shake the ship. They noticed it was Ching-Chong. Suddenly without warning, Ching-Chong put the ram to full and rammed so hard that Bob's spaceship went flying so far away that Ching-Chong couldn't see it. He started grinning. He knew it would be a while before they would come back

on track

-X-

"AAAAAH!!!!!!!!" yelled Bob and Joe as they tumbled backwards into darkness. They had been tumbling on and on for forever and then they stopped.

"Whoa" remarked Bob. It really was a pretty sight. There were pink, purple, blue, green, red, stars and even a black one. But there was more than that. There was a giant celestial object right in front of them. The portal to the Tredy Galaxy. It was massive---three times the size of the sun. There was a swirly design on the entrance. There was no sign that read, "Portal to the Tredy Galaxy" but Bob had a strong feeling that this was it. What if it wasn't? Well, they would just have to find out themselves.

"Let's go" Bob said to Joe, "It can't lead anywhere else."

Suddenly he stopped. Other than Trudy. There were rumors of a twin to the Tredy Galaxy, The Trudy Galaxy. While the Tredy Galaxy held the Stone of Ultimate Popularity, The Trudy Galaxy held the exact opposite. The Hata Stone. The Stone of

Ultimate Unpopularity. And they looked just the same. The portals looked the same too.

"Okay, we're going in" came Joe's voice from the steering wheel. Bobo strapped himself in his seat. "Ready Set, Off We Go" Joe said. They thrust themselves forward and entered he swirly portal. It felt like being stretched into a tunnel of rubber. It was awful. Bobo struggled in his seat. He couldn't think or move or do anything. He was paralyzed in fear. He just looked forward. Joe was sitting in his seat as well. His eyes were as big as pie plates and his face was pale. His mouth was in an O-shape his eyes filled with terror. Bobo felt sorry for him. After some 30 minutes they stopped.

"*That* was the scariest 30 minutes of my life" Joe said still shivering from the encounter.

"Mine too" Bob said.

"Hey man, we gotta run" Joe said. Fortunately, that was true. Bob wanted to get out of that place as soon as possible. But he also wondered if they went to the right portal. If they didn't and they took the wrong stone then they would become even more unpopular than they already were. And that would be bad. The only way to find out the difference between the twin galaxies was to go to the center. In the Tredy Galaxy, there was an orange star in the center. In the Trudy Galaxy, there was a green star in the center.

"Hey Joe, do you want to check if we're in the right galaxy?"

"Sure, but how exactly do we check?"

"Just fly right into the center. And if there is an orange star then we're in the right galaxy. If there is a green star then we're in the wrong galaxy."

"Okay, it will take about 7 days to reach the center. "

"You know what Joe. What if we're in the wrong galaxy and Ching-Chong is in the right one. He'll get the Stone first, right?"

"Yes he will unless he does something dumb and messes up. And it's a fat chance that he would mess up. So technically, he gets the stone and becomes popular if we're in the wrong galaxy because there's a portal only 90 trillion light years away from him!"

After their conversation Bob was trying to figure out a way to pass time. He forgot his PSP XL at home. He didn't even bring a book. The only thing left to do was try to sleep. He tried numerous times with no luck. Then finally, he collapsed from sheer exhaustion. When he woke up they were still flying except that they were on autopilot. Joe was asleep on his own bed. Bob couldn't blame him. Going through the portal must have exhausted him. Bob took hold of the wheel and then a while later, Joe woke up. His hair stuck up like suddenly out of nowhere BAM, a lightning bolt had struck him.

"Oh wow, we are really low on plutonium." Joe remarked glancing at the gauge. And he was right. Bob went to change the tank. Soon, they were on full. They continued onwards and noticed a catapult-like object floating around. It was tied to chains that were tied to planet of some sort.

After their experience with the Zunches, Bob was keen to stay away from non-human organic life forms.

They passed right by and they saw nothing except millions of stars. They found another queer-looking planet at was all tropical rainforests and hot weather. It was called Iapetuer. It never had any snow, just beautiful rays of sunshine. Bob and Joe were so tempted that they decided to dock on the planet for a day or two. They thought the inhabitants were going to be friendly but you will have to find out yourself if they were correct.

XI

"Hey Joe, don't you want to check out this place?"

"Sure"

"Let's go"

"Okay". So they flew closer and closer until they could land and they landed. It was the best place Bob had ever been since the Caribbean Islands. It smelled like a rainforest. There were even animals. There was one that looked like cross between a tiger and a monkey. It had a lion's head and a monkey's body. Its growl was in between a purr and a roar. It looked at them hungrily and then sulked into the shadows as if to say, "I would have eaten you but you aren't enough". They continued on and found a building the size of the Burj Khalifa.

"That must be the king's palace" Joe said.

"Yeah, I think you're right." Bob said. They entered without invitation. They just let themselves in. The creatures inside looked very queer. They had human heads and lion bodies. They talked in a musical voice that sounded beautiful. At first, they weren't noticed. Then someone yelled out a part of Fur Elise by Mozart. Supposedly, that basically meant "Intruder!" Immediately, armed guards threw them back on the wall.

"What are you doing here?" a guard said gruffly in English.

"We come in peace," said Bob.

"Sure," a guard said sarcastically. The guards had human heads and fish tails. Bob called them fishy boys. Anyway, the fishy boys took Bob and Joe to the kings throne area. It was huge. There were portraits hanging all around the room of what must have been the king's ancestors. At the very end of the room was a man sitting on a throne. He looked very grumpy. He was short and like the guards he had a human head and a fish tail. A cloak made of pure silver rested on his shoulder. His skin was mahogany and there was something like a cut on his hand that had healed but there was an orange ooze oozing out of the wound.

Blood thought Bob. The king appeared to be sleeping.

"King Cajun, we have brought prisoners." The king snapped awake. His face was twisted in anger.

"We come in peace" Bob said trying to lift the king's spirits. Unfortunately, that only made him madder.

"Throw them into the dungeon" the king spat.

"Wait I can explain" pleaded Bob.

"Why should I trust you?" the king sneered.

"You can search us but you will not find any weapons." Bob said confidently.

"Okay, search them" the king snarled. The guards searched them but as Bob promised they didn't find any weapons. When the king asked if they found any weapons the guards just shook their heads sadly. The king started to ease up a little.

"You have proven yourself innocent," the king said, "Come and join us for a welcoming feast." This time, the king himself escorted them. The room was empty. Then the king took a megaphone and basically yelled that there were guests. Nothing happened for a second, and then literally the whole entire palace came rushing into the dining hall. They rushed in and then they just stopped.

"Come on move it," the king screeched. Everybody unfroze and started laying plates and tablecloths and whatnot on the table. And then the chefs came out with the food. There was a blob of purple on Bob's plate.

"Mashed trewy root" the king said triumphantly. Bob tried not to make a face of disgust. He did *not* want to be thrown into the dungeon.

"Come on, have a bite," the king said enthusiastically. Bob plugged his nose and shoved a spoonful of the purple glop down his mouth. Surprisingly, it tasted like spicy hot chicken drumsticks. He wolfed the whole plate down. He could see that Joe liked it, too. They both asked for seconds but the king just laughed.

"Save some space for the entrée. That was just a healthy appetizer." He said.

"Healthy appetizer?" Bob said.

"Yep, mashed trewy root is the healthiest thing on Planet Iapetuer.

Bob was probably the happiest person ever after the king said that. Spicy hot chicken drumsticks=healthy. Bob could live on this planet for eternity and only eat chicken. He was so tempted that he thought about forgetting about their journey. But then he remembered that they might be in the wrong galaxy. He got up and said thank you and excuse us and with that they left.

XII

"Man, that was a pretty solid appetizer." Joe said.

"I agree," said Bob. They got in the spaceship and flew off. Bob felt nervous and scared. What if they were in the wrong galaxy and Ching-Chong was in the right galaxy? Then they would have to do a super boost, which could easily send them too far from the right portal. In order to do a super boost they would have to unlock a special fuel tank. In the tank they would have to put a bottle of Yurite. Yurite is the most flammable fuel known to mankind. Therefore, it burns so quickly it shoves the pistons into action so fast that it can propel anything 2000 times its maximum speed just enough to propel them from the center of the galaxy (if they were in the wrong one) to the other portal.

"We're going to be in the center in about half an hour," Joe said. Bob was half trembling half resisting the urge to jump up and down. After they passed by a bunch of orange stars Bob became more confident. "We are probably in the right one," he thought.

Suddenly an invisible force pulled them forward. They were sucked into a vortex. Inside they saw a greenish glow. They were in the center of the galaxy. The wrong galaxy.

"Joe do you think doing a super boost would get us out?

"Yes, but barely" came his reply, "And we only have one bottle" he added.

"You know that catapult-thingy that we saw earlier by Iapetuer? I think it's the exit out of this place." But Joe wasn't listening. He was already putting the Yurite in the special tank.

"You better strap yourself in your seat" Joe warned. Bob immediately agreed. He did *not* want to be thrown forward by a spaceship going 6,000 trillion light years a minute. That would certainly kill him. And he did not want to die in space. So he and Joe strapped themselves in their seats and Joe slowly and carefully pushed the thrust to full. Then, Bob and Joe were immediately thrown forward like bullet trains. They were on the fastest rollercoaster of all time. Bob screamed the loudest scream of his life and kept on screaming. He couldn't see move or do anything but scream. After a million years the ride on the fastest rollercoaster ended. Bob and Joe didn't unfreeze from their positions until 5 minutes after. Bob had stopped screaming. His throat hurt too much. After they unfroze Joe asked the question that Bob had already answered.

"How do we get out of here?" asked Joe.

"Well when you were putting the Yurite in its tank I said 'you know that catapult-thingy by Iapetuer? I think it's the exit out this place'"

"Let's go there," Joe said, not the least bit embarrassed that he didn't listen to Bob earlier. They flew to the catapult. It just floated there. Joe tried to figure out how it worked. He thought that all you had to do was to simply put your spaceship on the catapults bucket and then it would catapult you automatically, but it turns out that it needed a password. There was a statement that gave Joe an idea for the password. It read: The circumference: diameter is always constant.

"Hmmm… I wonder what that could be?" Joe said to himself, "Oh yeah it's pi." He put in the combination lock and directed Bob over to the catapult. Joe got back in the spaceship and they smugly fit inside the catapults bucket. Then they were catapulted. Bob had the same feeling that he had when they were being pulled into the portal. They landed about 10 minutes later in exactly the same place that they had been in when Ching-Chong had rammed them all the way to the other portal. They advanced to the portal. This portal looked exactly the same as the other one except it felt somehow warmer. They entered and this time instead of pain it felt like someone was giving him a massage. When they were in the Tredy Galaxy Joe tried to locate Ching-Chong and with luck they found out that he was only an hours flight away from them. But he was only feet away from the black hole. Joe put the thrust to full and they zoomed off into the space. Bob was too nervous to think about anything. They caught up with Ching-Chong and with a look of horror they saw that he was extracting the Zata

Stone from the Celebrium, the black hole where the Stone is located. Out of sheer desperation, Bob rammed into the arm of Ching-Chong's spaceship. The arm held for a minute but Bob knew that it was not meant to withstand that kind of force. The arm broke into a million pieces and the Zata Stone went flying into space. Bob's spaceship caught the Stone and it retracted into the spaceship. It was a magnificent piece of work. It was golden with flecks of silver in it.

"WE DID IT" he yelled out of happiness. But then, he saw Ching-

Chong advancing. He turned around and rammed Ching-Chong's spaceship in the back. The whole entire spacecraft lurched forward and then a miraculous thing happened. Ching-Chong hit the escape button on his spaceship and he was thrown out. He was wearing his space suit with an endless oxygen tank. He was gritting his teeth and shaking his fists angrily at Bob. Bob and Joe flew to Earth immediately. When they landed back in Bob's backyard they rushed in to the house so fast that they almost tripped over their own feet.

Bob's grandma hugged them so fiercely that Bob wondered why his ribs weren't broken. Anyway, they cracked open the stone using a hammer and found a glittering liquid inside.

"The Elixir Of Popularity." Bob said softly.

"And now for the mustard." Joe announced snapping Bob out of his trance. Bob's grandma brought the mustard and they each to half of the Elixir along with a squirtful of mustard. They poured the mixture into a glass each.

"Cheers" they said wincing a little. They gulped it down. And then it was over. It didn't even taste like anything.

"I just tasted air," remarked Bob.
"Me, too" Joe said. That night both boys had a sleepover. But really, all they did was sleep. When they woke they shoved some cereal down their throats threw some clothes and rushed to school. And then, they opened the door of the school to the cheering crowd of students who now knew that Bob was the popular one.

About The Author

Saaim Khan is a fifth grader at Cranbrook Schools in Bloomfield Hills, Michigan. His hobbies include hockey, soccer, robotics, swimming, playing the bassoon as well as the piano, and camping with his family. He represented Cranbrook Schools in Anaheim, California for a Robotics World Championship. Out of 72 teams, his team, the Bertie Bots, finished 22nd. He hopes to become a doctor when he grows up. He lives in Troy, Michigan and he also lives in

Sarnia, Ontario, Canada with his family.

www.ingramcontent.com/pod-product-compliance
Lightning Source LLC
Chambersburg PA
CBHW071807170526
45167CB00003B/1212